THE SCIENCE MUSEUM

The Solar System

Stephen Pizzey BSc

LONDON HER MAJESTY'S STATIONERY OFFICE

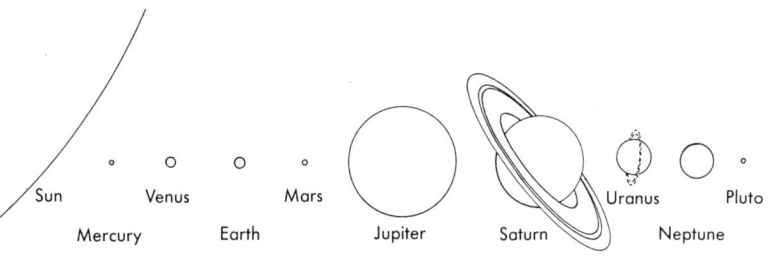

	Mean distance from Sun in million kilometres	Period of revolution	Period of rotation	Equatorial diameter in kilometres	Number of Satellites
Mercury	60	88 days	59 days	4 880	–
Venus	108	225 days	243 days	12 104	–
Earth	150	365 days	24 hrs	12 756	1
Mars	228	687 days	24½ hrs	6 787	2
Jupiter	778	11.9 yrs	10 hrs	142 800	13
Saturn	1 427	29.5 yrs	10¼ hrs	120 000	10
Uranus	2 870	84.0 yrs	11 hrs	51 800	5
Neptune	4 497	164.8 yrs	16 hrs	49 500	2
Pluto	5 900	247.7 yrs	6¼ days	6 000 (?)	–

1 The Planets compared.

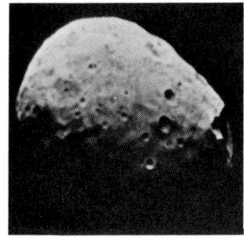 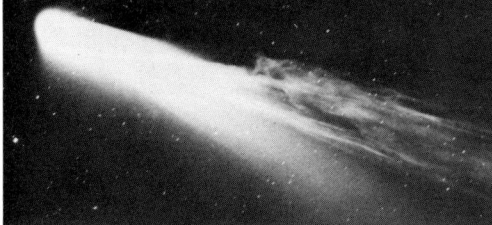

2,3 The Sun and Planets are not the only residents of our Solar System – asteroids and comets also orbit the Sun. Shown above is Phobos, an asteroid-like moon of Mars, and the comet Mrkos.

The Solar System

We are travelling through space on the surface of a planet which is itself revolving, along with eight other planets, around an average size star, the Sun. Our planet has one moon which we now know differs in composition from the Earth. The inner planets, Mercury, Venus, Earth and Mars are rocky in nature whereas the giant planets Jupiter and Saturn are mainly liquid and less dense than the others. Little is known of the outer planets Uranus, Neptune and Pluto due to their great distance from Earth, but it is thought that Uranus and Neptune are similar to the gaseous planets Jupiter and Saturn. Pluto is a small planet and probably rocky. In the 1980s Neptune will become the farthest planet from the sun as Pluto's orbit takes it inside that of Neptune.

The Sun and planets with their attendant moons are not the only residents of our solar system; there are also asteroids, large fragments of rock which lie mainly in a belt between the orbits of Mars and Jupiter. Some of the asteroids have orbits which take them close to the Sun and occasionally close to Earth. Collisions with such objects resulted in large impact craters of the type seen on the Moon. Small pieces of debris on a collision course with Earth show themselves as 'shooting stars' as they enter the Earth's atmosphere and burn up due to the enormous friction. Larger pieces occasionally reach the surface as meteorites and until the Apollo missions returned samples from the Moon these meteorites were the only examples we had of material from elsewhere in the solar system.

How then did the solar system evolve into the form we know it today? There have been, and still are, many explanations but any theory must stand up to mathematical scrutiny and account for the rotation of the Sun and planets, their composition, the fact that their orbits lie more or less in a plane, and the existence of comets. It is now generally agreed that the Sun and planets were formed from a rotating disc of dust and gas, but the details of exactly how this occurred are hotly debated. The Great Nebula in the constellation of Orion (fig 4) is a mass of glowing gas and dust which can be seen with the aid of binoculars and appears as a small 'puff' of luminous smoke. Stars are forming in this nebula and it was within such a cloud that the sun and planets formed 4,600,000,000 years ago.

It is thought that as the central gas cloud contracted, and the 'solar nebula' that was to become the Sun started to form, dust particles collected together to form clumps or grains which in turn collected to form millions of larger bodies in a disc around the solar nebula. These in turn became grouped together by gravitational attraction and collision to form large planet sized bodies which became heated by a combination of the crushing effect of their own gravity, the bombardment of the surface by the infall of more material, and natural radioactivity. The whole mass of each planet melted and later cooled. The solar nebula had meanwhile contracted and the Sun began to form and become hot.

Some planets, like Mercury, still bear the scars of impacts on the surface whereas the Earth has lost most of its record of early events as a result of geological activity, which continues today.

Comets travel from the far reaches of the solar system and pass close by the Sun. They are thought to be a collection of dust, ice and gas and have been described as 'dirty snowballs'. Comets were known as 'hairy' or 'bearded' stars in ancient times and it was popularly believed that they foretold disaster. Edmund Halley (1656–1742) disposed of this belief by predicting the return of a comet in 1758. His prediction was correct, although Halley never lived to see it.

The solar system is pervaded by a 'solar wind' of electrically charged particles which constantly stream out from the Sun. The Aurora or 'Northern Lights' result from large numbers of these particles becoming trapped by the Earth's magnetic field and guided into the upper atmosphere.

4

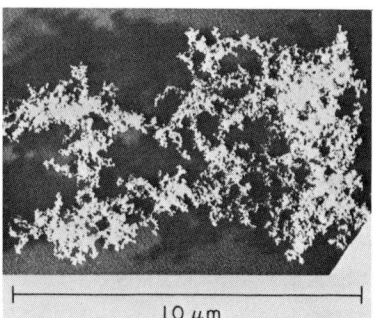

5

4 The Great Nebula in Orion: Stars and Planets are forming in this distant cloud of dust and gas. Our own Solar System formed within such a cloud 4·6 thousand million years ago.

5 A 'fluffy' grain captured by a high altitude balloon and thought to be of interplanetary origin. The planets started to form by the progressive accumulation of small grains of this type.

The Sun

The Sun is our nearest star and our lives depend on the constant outpouring of its heat and light. Our climate is driven by energy from the Sun and all our fuels, except nuclear fuel, have depended on solar energy for their formation. In burning coal, gas, oil or wood we are unlocking energy from the Sun. How then does the Sun generate its energy? The answer was for many years a mystery. A giant ball of coal the size of the Sun would only burn for a few thousand years and the Sun has clearly been with us for longer than that. The answer is impressively demonstrated in that awesome symbol of the twentieth century, the hydrogen bomb. Deep inside the Sun's core 564,000,000 tons of hydrogen are converted into 560,000,000 tons of helium every second. The lost 4,000,000 tons are converted into energy which eventually reaches the surface as heat and light.

In 1973 and 1974 scientists manning the solar observatory aboard the 'Skylab' space station studied the Sun as never before. Eight and a half months of continuous observation from this vantage point 500 kilometres above the Earth showed the Sun to be surrounded by a seething atmosphere of gas at millions of °C. Much has been learned from these and other recent observations from Orbiting Solar Observatory (OSO) and Copernicus satellites. The results show the Sun to be in a state of constant turmoil. In addition to light and heat, charged particles stream out to form a continuous solar wind. Solar eruptions on an immense scale such as those shown in the photographs, can disrupt radio communications on Earth as large numbers of high energy particles collide with the Earth's upper atmosphere.

The 'Skylab' space station was the largest manned spacecraft to be placed in orbit around the Earth and it afforded the crew a spacious shirt-sleeve environment in which to conduct studies of the Earth and Sun. Skylab was by far one of the most significant attempts at living and working in space. Instruments capable of observing the Sun simultaneously in X-rays, ultra

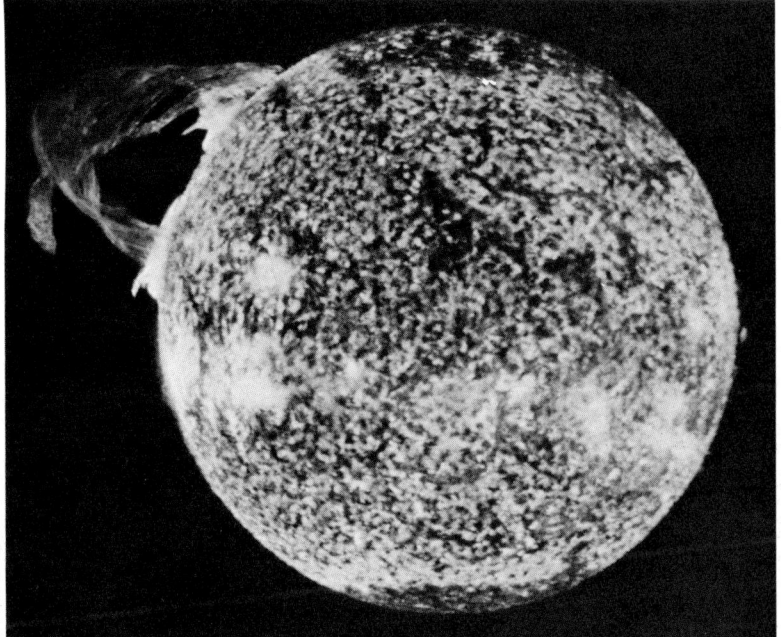

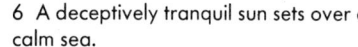

6 A deceptively tranquil sun sets over a calm sea.

7 Solar eruptions on an immense scale can disrupt radio communications on Earth as high energy particles blown out from the Sun collide with our upper atmosphere. This was taken from the 'Skylab' space station in ultra-violet light.

8 The 'Skylab' space station. The tower at one end is the solar observatory.

9 A member of the crew aboard 'Skylab' at the solar observatory control panel.

(5)

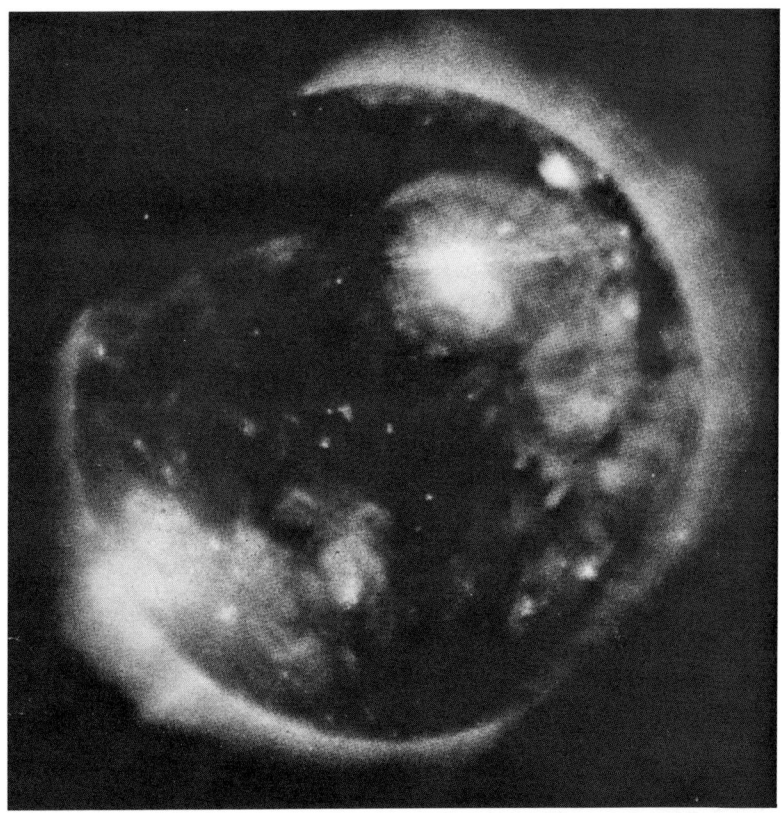

violet and visible light were housed in the Apollo telescope mount, the tower-like structures at one end of Skylab. The space station was launched with no-one aboard on 14 May 1973. Soon after lift-off, a protective heat shield and a solar panel were torn away leaving a second solar panel jammed and unable to open in orbit. Skylab began to overheat and it would have only been a matter of days before the space station became a useless shell. A makeshift parasol was hastily prepared on Earth to counteract the overheating and the crew practised freeing the locked solar panel on a full scale mock-up underwater to simulate weightless conditions. The crew were then sent into orbit to repair the damage. Eventually they were successful in freeing the panel and the parasol meanwhile shaded the exposed section. The interior of Skylab began to cool and the space station became operational.

Although the exterior of the Sun can be examined in detail, there is no such direct method of observing the interior. One experiment designed to probe the interior is housed deep in a gold mine in South Dakota USA and intended to detect high energy sub-atomic particles called neutrinos. These particles travel direct from the Sun's interior and are capable of passing right through the Earth. Neutrinos are a product of nuclear reactions taking place in the Sun's core. The earth and rock above the detector filters out low energy neutrinos from other sources.

In 10 years of operation, far fewer neutrinos have been detected than predicted by theory. This negative result is of major concern to astronomers, since the neutrinos *should* be produced if their understanding of nuclear processes is correct.

A second method of studying the Sun's interior relies on measuring the diameter of the Sun. The Sun is shaking like a jelly and variations in the diameter indicate how vibrations travel through the core, giving a clue to its nature.

10 Studying the Sun's exterior:
The turbulent nature of the Sun's atmosphere is revealed by this X-ray photograph from Skylab.

11 Studying the Sun's interior:
There is no *direct* way of studying the interior of the Sun. The experiment shown here is housed in a mine in South Dakota USA and designed to detect sub atomic particles which travel direct from the Sun's core.

Mercury

Before the voyage of Mariner 10 to within a few hundred kilometres of Mercury, only guesses could be made as to the detailed nature of its surface. It was known that Mercury reflected light from the Sun and radar pulses from Earth in much the same way as the Moon would at that distance, but that the interior must be iron rich and more akin to the Earth than the Moon. Before the days of space exploration there was disagreement among leading authorities as to whether Mercury had an atmosphere and whether the planet rotated such that one face was always turned towards the Sun.

The remarkable photographs from Mariner 10 show that the surface of Mercury in fact bears an uncanny resemblance to the Moon. Further results show that the planet has no atmosphere save for a few particles from the Sun and that it rotates slowly on its axis once in 58·65 days while it revolves around the Sun every 88 days. Mercury's rotation and revolution time exhibits an effect known as 'spin orbit coupling' as the two are in exact ratio of 2:3. Although strikingly Moonlike in appearance there are detailed differences between Mercury and the Moon such as cliffs on Mercury which run for hundreds of kilometres. The distribution of craters gives geologists on Earth a clue to Mercury's geological history. Mariner 10 also discovered that Mercury has a weak magnetic field.

There are surprising geological similarities between Mercury and the Moon, considering the great distance between them and the question arises whether Mercury, the Moon, and presumably the Earth were bombarded by the same groups of asteroid-like bodies in the early history of the solar system, or whether the planets underwent their own separate periods of bombardment.

The Mariner 10 spacecraft used the gravitational pull of the planet Venus to swing the craft into a favourable orbit around the Sun to enable close-up photography of Mercury. This 'sling-shot' technique of using another planet's gravity to accelerate a spacecraft into a new trajectory has become a standard practice for multi-planetary exploration.

12 This photomosaic of Mercury, the nearest planet to the Sun, was compiled from pictures taken by TV cameras aboard the Mariner 10 spacecraft and shows the cratered, moonlike appearance of the planet's surface.

Venus

The Mariner 10 spacecraft which successfully photographed the surface of Mercury in such detail obtained no such photographs of Venus. The entire surface of Venus is obscured by clouds almost 50 kilometres thick and thought to be composed of sulphuric acid droplets with traces of hydrochloric and hydrofluoric acid. The surface of the planet is at a temperature of around 470°C, almost the temperature at which objects glow red hot.

The atmospheric pressure at the surface is equivalent to that in the Earth's oceans at a depth of 1 kilometre and is due to an extensive atmosphere of carbon dioxide. The planet Venus was once regarded as a sister planet of Earth in terms of size and density but why is the surface of Venus so hot and so unlike Earth? The high temperature may be due to the 'greenhouse effect' whereby incoming solar radiation can penetrate the atmosphere but outgoing thermal radiation from the surface is trapped, thereby heating the surface layers.

Why is the atmosphere of Venus so massive compared to the Earth's? A simple answer is that if the surface of the Earth had been as hot as Venus then carbon dioxide would have been liberated from the rocks and Earth too would have such an atmosphere. But this is really a

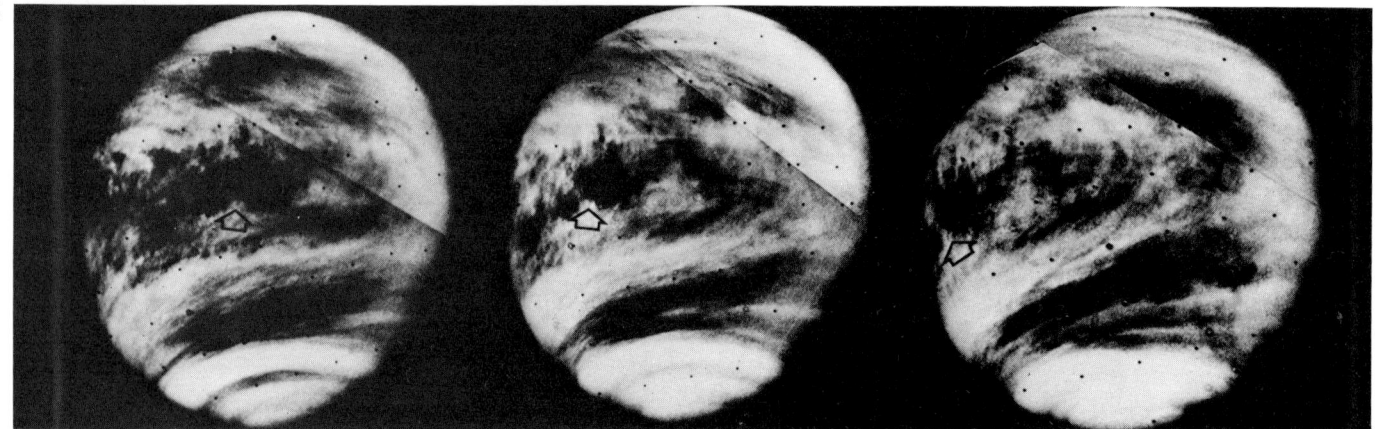

'chicken and egg' argument since the high temperature may have been due to the atmosphere in the first place.

The answer to how the greenhouse effect started may lie in the fact that Venus, being nearer the Sun, receives nearly twice as much solar energy as the Earth and this may have raised the surface temperature sufficiently to release just enough carbon dioxide to trigger the effect. This would result in a temperature rise and release more carbon dioxide and so on. An intriguing question is that if Venus was Earth's sister planet how did it come to lose all its water, and why is there no water vapour in the atmosphere?

Although the cloud tops reflect most of the sunlight, approximately 1 per cent filters through so that the light level at the surface during the day is equivalent to a dull day on Earth when dark storm clouds are brewing.

If the study of Mercury from Mariner 10 was a geologist's delight, then Venus must be a meteorologist's delight in that both it, and its atmosphere, represent a simplified weather system to test theories and models of aspects of Earth's own climate.

But what is known of the surface of Venus? In 1975 the Russian spacecraft Venera 9 and 10 successfully penetrated the thick cloud layer and survived the high surface temperature long enough to transmit pictures from the surface showing stones measuring up to 1 metre in diameter on the dark Venusian 'soil'. Further information has been obtained from the Arecibo Radio telescope by using the telescope as a giant radar. Radar waves were transmitted from the telescope and bounced back from the surface of the planet. Pictures of the surface have been processed from the return signal since radar waves, unlike light waves, are not scattered by the clouds. These radar pictures show bright areas due to lava flows and show Venus to have large diameter low walled craters with no large mountains or deep valleys. Radar measurements also show that Venus rotates about its axis once every 243 days whereas Mariner 10 photographs show levels in the vicinity of the stratosphere, about 100 kilometres above the suface, swirl around the planet every 4 days.

15

16

15 The rocky surface of Venus as photographed by the Russian Venera 9 lander. The surface temperature on Venus is over 470° C.

16 The character shown sitting on the base of Venera 9 with the surface of Venus in the background appeared on a Russian greeting card sent to fellow scientists engaged on planetary research.

13 Photomosaic of Venus taken from Mariner 10 in ultra-violet light, showing cloud cover.

14 Photographs taken at seven hour intervals showing rapidly rotating cloud features above the slowly rotating planet.

The Earth and Moon

We are all familiar with the planet Earth. A visiting space probe circling our planet would show clouds in the atmosphere, oceans, mountains, deserts, icecaps and vegetation. Detailed photographs would pick out signs of cultivation, urban areas and pollution. A lander touching down on our planet would probably be confronted by many strange forms of life ranging from bacteria to large mammals. The Alien Mission Controllers would watch in dismay as their lander was carried off by some strange two-legged creatures.

Whereas the craters and lava flows on the surface of the Moon are a timeless record of the Moon's early history, no such complete record survives on Earth. The Earth is geologically active, continents drift slowly over its surface on giant rafts floating on the underlying mantle, volcanic action continues today and wind and water constantly erode the surface features. Echoes recorded from underground nuclear explosions and earthquakes have shown that the Earth possesses a solid inner core, which is surrounded by a liquid outer core. Our planet has a magnetic field which is thought to originate from a combination of the Earth's rotation and the electrically conductive liquid core creating a dynamo. Rock samples have shown that the magnetic field has reversed many times in the Earth's geological history.

This planet is being actively studied from the ground, from under the sea and out in space. Many of the thousands of satellites circling above are committed to the study of the Earth and its resources.

One of the earliest discoveries of the space age was made in 1960 by 'Explorer', the first American satellite. Explorer detected two doughnut shaped belts of radiation lying outside the Earth's atmosphere. These 'Van Allen' belts, as they are now named, are made up of charged particles from the 'solar wind' which have become trapped in the Earth's magnetic field.

The Apollo Moon landing in 1969 provided scientists with a first-hand look at a planet-like body other than Earth. It is on the Moon's ancient surface that many of the clues to the early history of the solar system lie. The

17 An alien visitor to our planet would find it one of the most varied objects in the solar system, with ice caps, oceans, an atmosphere, continents and, above all, life.

18 The Earth's magnetic field forms a magnetic 'envelope' around the planet and traps charged particles in a doughnut shaped ring known as the Van Allen Belts.

(10)

Apollo manned missions to the moon

	Date of launch	Astronauts	Landing site	Comments
Apollo 11	11.7.69	N Armstrong E Aldrin M Collins	Sea of Tranquillity	First man landing on the Moon
Apollo 12	14.11.69	C Conrad A Bean R Gordon	Ocean of Storms	Astronauts visited Surveyor
Apollo 13	11.4.70	J Lovell F Haise J Swigert	–	Mission aborted before landing
Apollo 14	31.1.71	A Shepard E Mitchell S Roosa	Fra Mauro	Used hand buggy to carry equipment
Apollo 15	26.7.71	A Scott A Worden J Irwin	Hadley Apennine region	First use of lunar rover
Apollo 16	16.4.72	J Young T Mattingly C Duke	Déscartes	Similar to Apollo 15
Apollo 17	7.12.72	E Cernan R Evans H Schmitt	Taurus Littrow	Last manned lunar mission

oldest lunar rocks returned by the astronauts were formed over 4,000,000,000 years ago. 4,500,000,000 years is the age of the oldest meteorites yet found, whereas on Earth the rocks are much younger. Although the Earth and Moon differ in composition, oxygen isotope studies of their rocks indicate that they were formed in the same region of the solar system. This finding rules out the suggestion that the Moon was captured by Earth from some orbit nearer the other planets. The most favoured explanation seems to be that the Moon was formed from a collection of material in Earth orbit.

19 Clues to the formation of the Solar System lie on the ancient lifeless surface of the Moon. Shown here is astronaut Scott and the Lunar Rover (Apollo 15).

20 By reflecting laser beams back to Earth from reflectors left on the Moon, scientists are able to detect movements of the Earth's crust by using the Moon as a reference. The technique is known as 'Lunar Laser Ranging'.

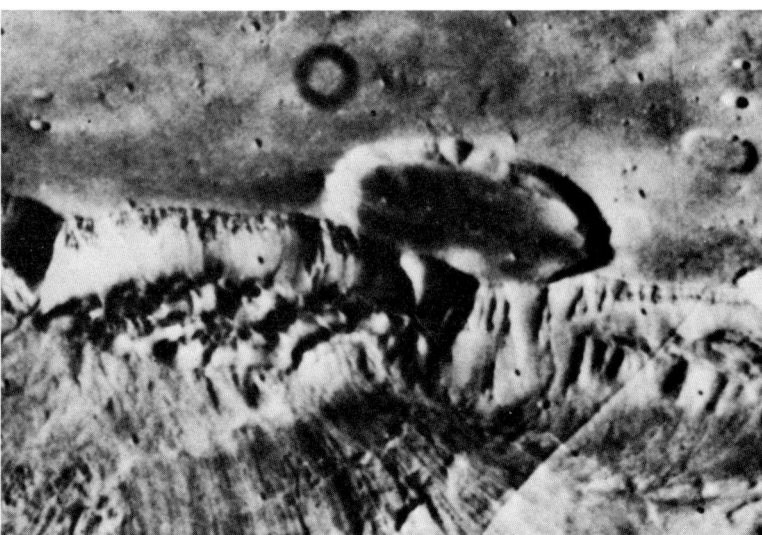

21

22

21 The 2 km high side of part of the Valles Marineris, the Grand Canyon of Mars, is shown in this photograph from the Viking 1 orbiter.

22 Rock strewn terrain around the Viking 2 landing site.

opposite
23 A cloud of gas streams out from the Sun at a speed of 6500 kilometres per minute. The Earth would appear only 5 milimetres in diameter on the same scale. (Electronically processed XUV photograph from the US Naval Research Laboratory experiment aboard Skylab.)

Mars

In the minds of many people, Mars, the red planet, has always been a favoured stage for the science fiction heroes, advanced civilisations and a platform from which to invade a helpless Earth. In 1937 the H G Wells classic *War of the Worlds* was broadcast on American radio by Orson Welles and narrated as if the invasion of Earth by Martians was actually taking place. Thousands of people took to their cars and fled in the face of a fictitious Martian invasion – a tribute to Orson Welles' performance and people's credulity.

The recent Viking unmanned missions to Mars have shown us that Mars is indeed a red planet, the sky is pink, and the powdery wind blown soil is red. Mars is a world of towering volcanoes and deep canyons, and water once flowed in torrents over the now dry river beds. The thin atmosphere supports cloud formations and the ice caps are now shown to be water ice and not solid carbon dioxide as previously thought. The temperature on Mars lies between −140°C and −30°C, which means that the highest temperatures on Mars correspond to the lowest temperatures on Earth.

The Viking 1 and Viking 2 spacecraft went into an orbit around the planet in the July and August of 1976. Provisional landing sites for the lander part of the spacecraft had been previously selected from pictures from the earlier Mariner 9 mission in 1972.

A flat area the size of Yorkshire would have provided the safest site for the craft as the Viking Lander has a ground clearance of only 220 millimetres and a large landing area was required to allow for errors. No such area could be found, so choosing a site became a task of choosing the least hazardous of a number of potentially hazardous sites. The main instruments for assessing large-scale features on the surface, once the spacecraft had arrived, were the orbiter cameras which showed surface details down to a scale of 100 metres. A rock of, say, 50 metres on the surface would not be discernible from the orbiter camera and yet could cause serious damage to the lander. An indication of the nature of the surface was obtained from thermal measurements and

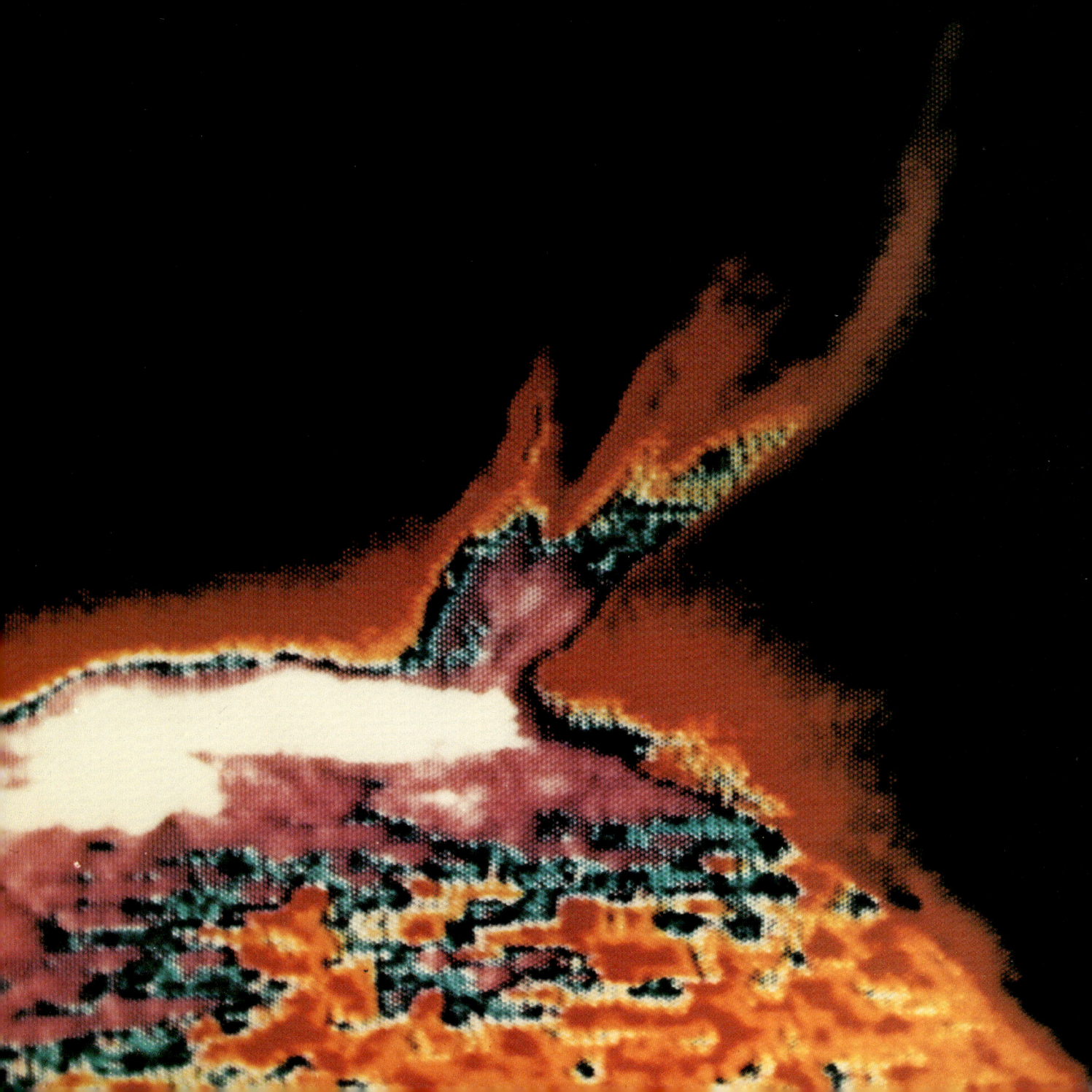

Mars – The Red Planet

24 From Viking 1 at the Chryse Planitia landing site.
25 From Viking 2 at Utopia Planitia.
26 Valles Marineres Canyon from the Viking orbiter, the sides rise to over 5 kilometres.
27 Mars as seen by the approaching Viking spacecraft.

following page
28 The giant planet Jupiter as seen by the Pioneer 11 spacecraft, showing parallel bands of cloud in the planet's atmosphere. The Great Red Spot is clearly visible and is also a cloud feature.

(16)

from radar signals sent from Earth and bounced back from the Martian surface. The prime landing site was originally intended to be in the Chryse plain, 19·5°N, 3°S W; to the dismay of the project scientists the orbiter photographs showed that meandering channels ran across the site and so project officials decided to try an alternative site.

The lander could not be guided down because the signals would have taken 18 minutes to cross the 300,000,000 kilometres separating Earth and Mars and instructions to change course would take a further 18 minutes to reach the lander – by which time it would have landed. Eventually a site was selected and the lander separated from the orbiter to begin its descent to an unknown fate either as the first fully operational laboratory on Mars or as a tangled heap of metal.

The lander touched down safely, although the mission controllers were not to know for an agonising 18 minutes until the signal reached Earth. Meanwhile the orbiter continued to circle the planet. Once on the surface, the cameras on board the lander scanned the scene; there were no plants, no trees and no Martians, instead a desert-like scene with rocks all round. The first photograph from the landers showed the number 3 footpad on the Martian soil, the second photograph showed a panoramic view of the surrounding terrain. Viking 2 showed similar views from a different site. The landers carried instruments which acted as weather stations on Mars collecting and transmitting data on wind speed, temperature and atmospheric pressure. A typical daily weather report for late Martian summer from the lander site would show a maximum temperature of –30°C at 3.30 in the afternoon with a maximum windspeed of 8 metres per second with gusts of up to 15 metres per second.

The lander also carried a miniaturised automatic laboratory designed to test samples of soil scooped from the surface of the planet for signs of life. Soil was collected from the surface by a sampling arm and emptied into a hopper on the lander. The sample of soil was then added to each of three experiments. These were the Carbon Assimilation or Pyrolytic Release Experiment which tests for a photosynthesis-like process by subjecting the sample to artificial

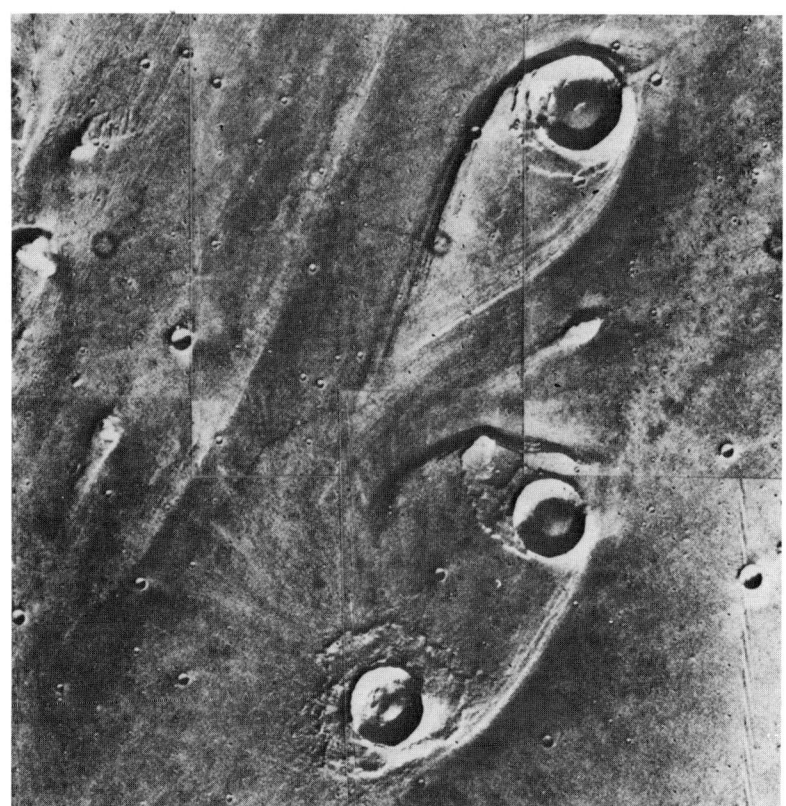

29 Water once flowed on Mars as shown by this photograph of tear-drop shaped islands from the Viking 1 orbiter.

30 Weather reports from Mars come from the meteorological booms on board the Viking landers. Shown here is the Viking 1 boom against the Martian horizon.

31 Foot pad of the Viking lander on the Martian surface.

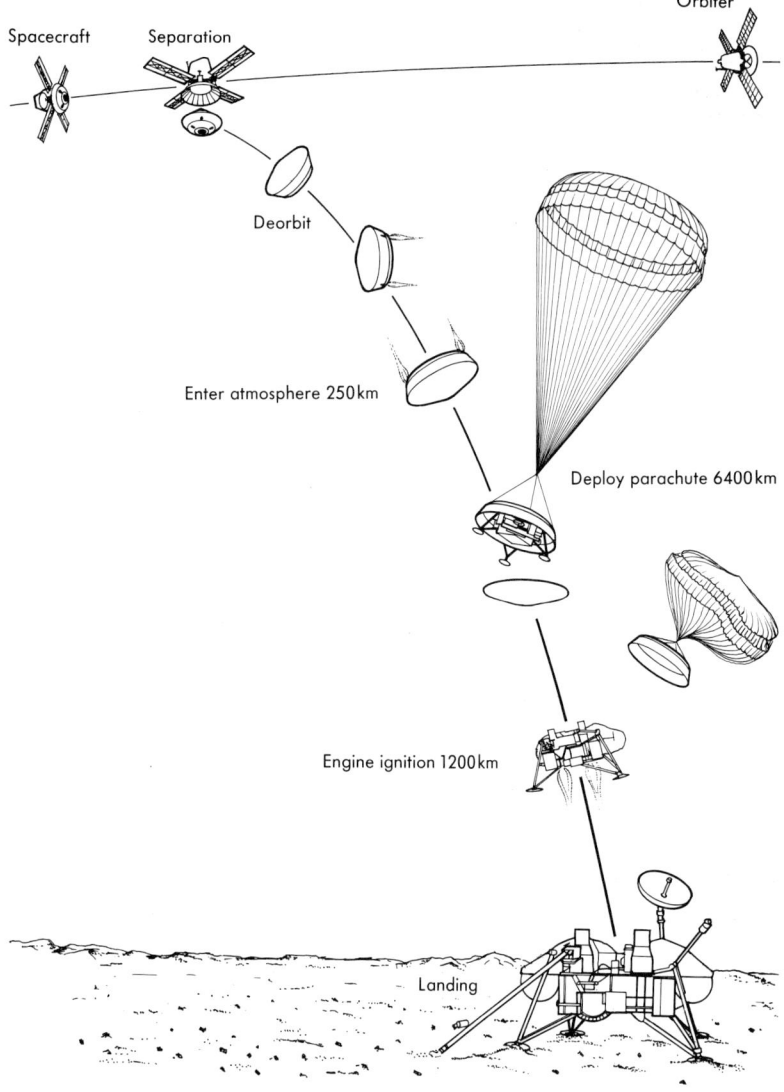

32 The landing sequence: the lander descends to the Martian surface as the orbiter continues around the planet. During the long journey to Mars, the lander and orbiter are coupled together.

33 Mars' two tiny moons, Deimos (a) and Phobos (b), measure only 12 km and 25 km across. Photos are from the Viking 1 orbiter at ranges of 3300 km and 500 km.

sunlight and testing whether the soil takes in carbon dioxide. Radioactive labelled carbon dioxide is used and the sample is then heated to see if there has been assimilation of the carbon dioxide by detecting radioactive carbon-14 in the sample.

The second experiment, termed the 'Labelled Release Experiment' in effect feeds the soil with a nutrient soup labelled again with radioactive carbon-14. During the experiment the gas above the sample is tested for carbon-14 to see if something in the soil is releasing gas containing carbon as a result of metabolism.

The third experiment, the 'Gas Exchange Experiment' tests the sample by adding nutrient and then testing the change in composition of the gas in the test chamber to see if there are any processes taking place which either take in or liberate gases.

Many assumptions had to be made when designing these experiments, two of which were that Martian life would be based on molecules containing carbon and that biochemical reactions of Martian micro-organisms would require water vapour, as on Earth. The Martian soil was found to be active but the results were inconclusive. The 'Pyrolytic Release' experiments showed that synthesis occurred in the sample but that it was weaker than in biologically active soil found on Earth. The 'Labelled Release' experiment also showed the soil to be active as large amounts of gas were released which could be due to biological activity. The third experiment again showed the soil to be active but the results could be explained either by a chemical or biological process.

So, at first sight, these results seem to indicate the presence of life on Mars. The note of caution is sounded by the results of a fourth experimental package designed to perform a chemical analysis of the soil. No organic compounds were found in the Martian soil, yet organic compounds should have been present if organisms live or had lived there. When tested in the Antarctic on Earth this experiment detected over 20 organic compounds in the soil. The consensus of opinion is that the results are due to some complicated, and as yet unexplained, chemistry.

Mars has two small moons, Phobos and Deimos, which were first detected in 1877 as two faint specks

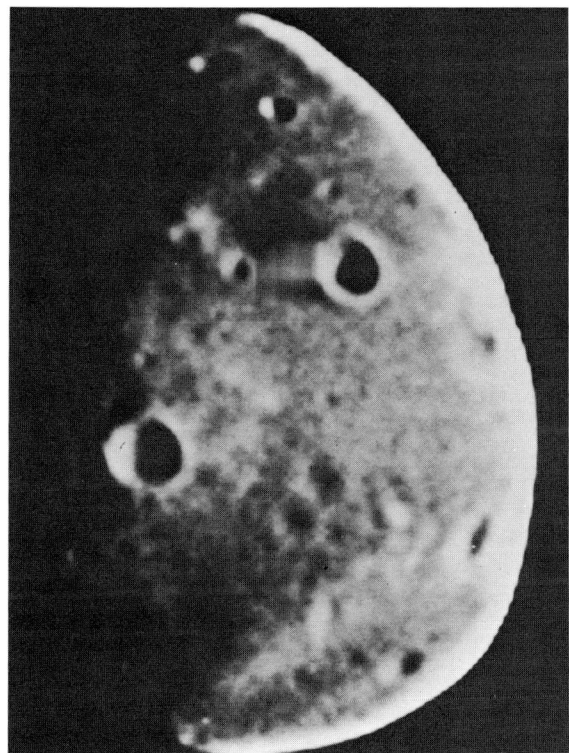
33a

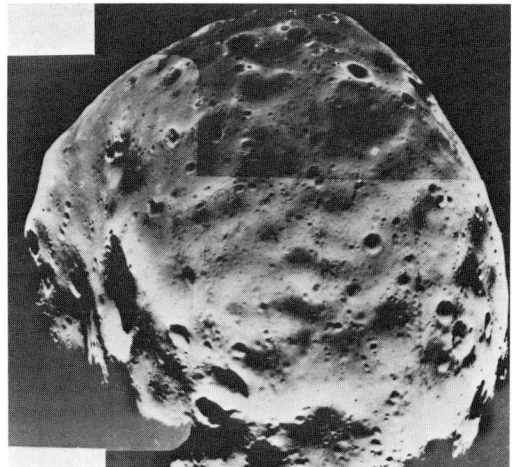
33b

34 The Viking lander

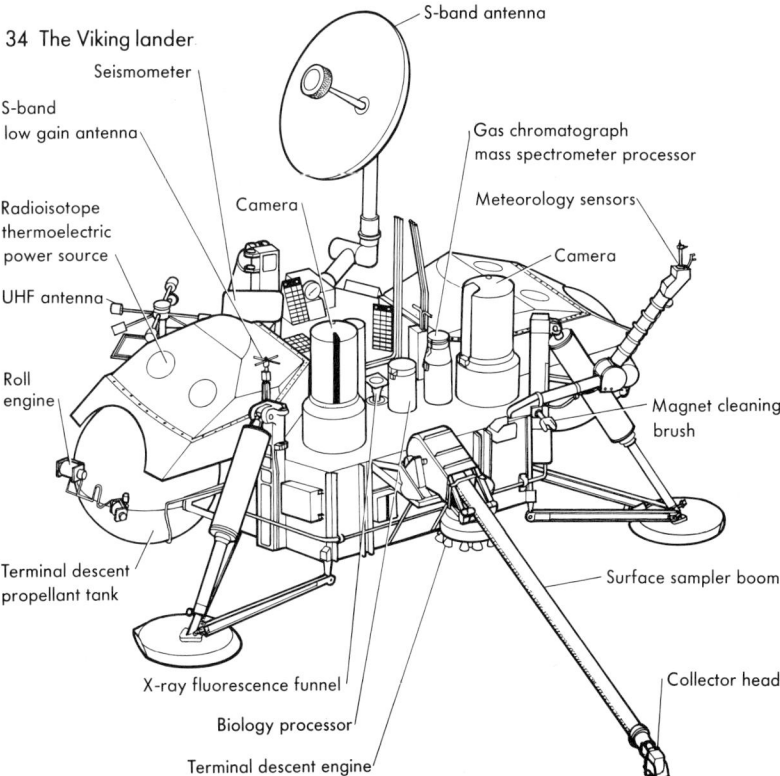

near the planet. The moons have been photographed by the Mariner 9 spacecraft, and in detail by the Viking orbiter. They appear to be cratered rugby-ball shaped fragments of rock and are grey in colour. Phobos is approximately 25 kilometres in diameter and Deimos is approximately 12 kilometres. Calculations based on measurements of Phobos' orbit show that in 100,000,000 years time Phobos will collide with Mars. Photographs from the Viking orbiter represent the first detailed inspection of an asteroid-like body. The study of Phobos and Deimos may provide further pieces in the jigsaw pictures of how the planets formed 4,600,000,000 years ago.

(19)

Jupiter

The planet Jupiter is a giant in the solar system and quite unlike any of the inner planets in that it is liquid and has 13 moons. When viewed from Earth, through even a modest telescope, various features are readily apparent. The four large moons, themselves the size of small planets, are clearly visible just as they were to Galileo in 1610. Flattening can be seen at the poles due to Jupiter's rapid rotation, and light and dark bands running parallel to the equator may also be discerned. The bands of cloud which totally cover the planet show a pattern of alternate belts and zones (see fig. 36). The light-coloured zones are higher and cooler than the coloured belts. The cause of the coloration is as yet unknown. The images sent back to Earth from Pioneer 10 and in greater detail from Pioneer 11 show that the bands are less permanent at high latitudes and break down into whorls, streaks and generally irregular features near the poles. Detailed study of weather patterns on Jupiter from the Pioneer data will enable meteorologists to test theories on Earth's complicated atmospheric circulation by using Jupiter as a simplified model. Study of meteorology of the planets will contribute to the understanding of aspects of our own climate.

There is a 'Great Red Spot' which has for centuries been a conspicuous feature in Jupiter's atmosphere and many suggestions as to its nature have been advanced over the years. The region measures 30,000 kilometres across and may possibly be a permanent hurricane. Computer models based on the Pioneer data suggest that such a large storm feature could sustain itself and 'freewheel' around the planet indefinitely. The observation of a 'Little Red Spot' (similar in appearance but smaller) on the Pioneer 10 images adds weight to this theory.

Jupiter is enclosed in a magnetic envelope in a similar way to Earth due to the magnetic field generated by the rotation of the planet. In the case of Jupiter, liquid metallic hydrogen, unknown on Earth, is thought to be formed within the planet due to the combination of high

35 An artist's impression of the Pioneer 10 spacecraft as it flew by Jupiter in 1973. The radio dish measures 3 metres across and the finned structures on booms are small radioisotope nuclear generators which supply electrical power for the spacecraft.

36 Detailed photographs of Jupiter from Pioneer 11 as it flew by in December 1974. Pioneer 11 is due to arrive at Saturn in 1979.

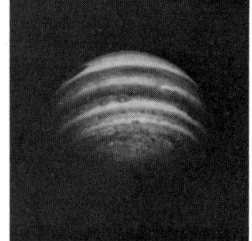

pressure and high temperature. This material is electrically conducting and would create a dynamo effect as the planet rotates, thus generating the magnetic field. Jupiter's magnetic field strength is 17,000 times that of the Earth's and traps particles which form radiation belts around the planet. The intensity of the radiation is equivalent to that from a nuclear explosion. Three of Jupiter's big moons move through the heart of the radiation belts, and one, Io, is responsible for massive radio blasts as 10,000,000,000,000 watts is dissipated in current flows between Io and the planet.

The chemical composition of Jupiter is similar to the Sun and the formation of Jupiter and its moons is thought to be akin to the formation of the Sun and planets. If Jupiter had been more massive then the heat generated during its contraction would have been sufficient to initiate nuclear fusion and Jupiter would have shone as a star. Jupiter still has a hot interior and emits twice as much heat as it receives from the Sun.

The two Pioneer spacecraft returned data which have given scientists a much clearer understanding of the planet and inevitably raised more questions; questions which may be answered by the future Voyager spacecraft, due to encounter Jupiter and its moons in 1979 and Saturn in 1981. Pioneer 11 is meanwhile on course for Saturn and due to fly close by in 1979.

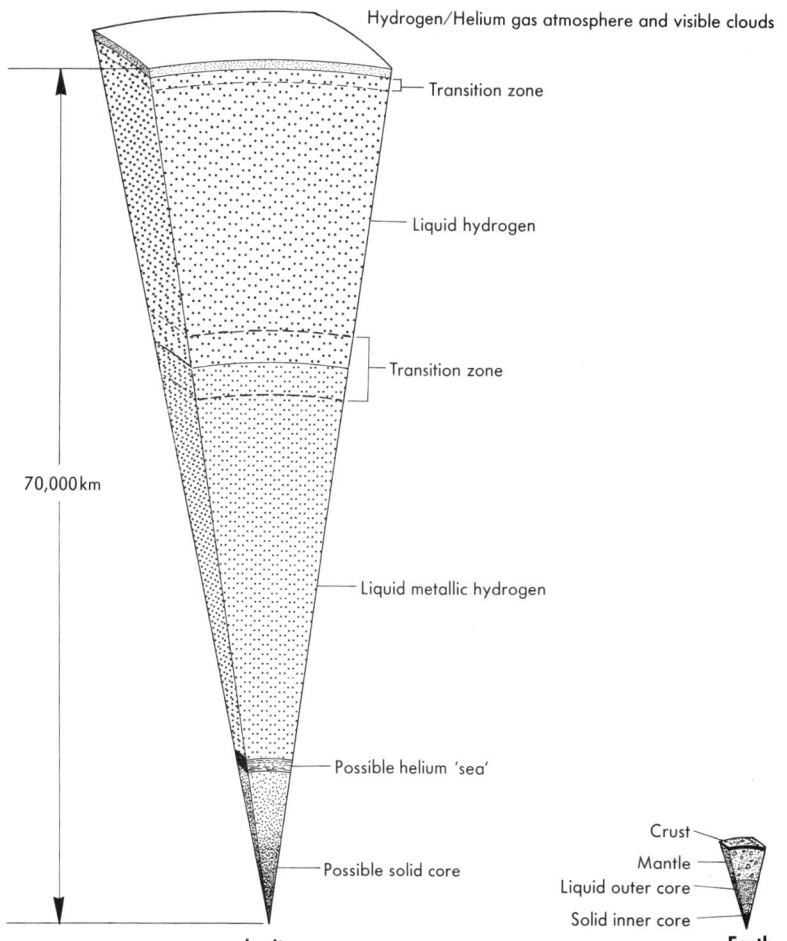

37 Jupiter and Earth compared. Jupiter is a giant liquid planet and over ten times the diameter of Earth. Like Earth, Jupiter has a magnetic field.

Saturn and the outer planets

Saturn lies 1,400,000,000 kilometres from the Sun and, like Jupiter, is a giant, rapidly-rotating liquid planet. Unlike Jupiter, all the information up to 1979 when Pioneer 11 approaches Saturn, will have been gathered from Earth-based observations. Saturn rings present the most visually distinctive feature of the planet. Radar reflections returned from the rings indicate that they are made up of particles of ice measuring between 40 and 300 millimetres across. The spectrum of sunlight reflected from the rings was found to match water ice at very low temperature and also showed that the rings rotate at different rates. One theory suggests that the rings are a group of particles which never collected together to form a compact moon.

Saturn has 10 moons of which Titan is the largest and almost the size of the planet Mars. Titan also has an atmosphere which is denser than that of Mars. Close inspection of Titan will be one of the objectives of the Voyager mission to Jupiter, Saturn and Uranus. (The Mariner class spacecraft has been named Voyager). Photographs of Saturn show a flattening at the poles and bands running parallel to the equator which are thought to be cloud tops in a deep atmosphere.

The three outer planets Uranus, Neptune and Pluto are too far from Earth for much to be known about them. Uranus is unusual in that it is spinning about an axis in the plane of its orbit, like a gyroscope on its side. Recent data obtained as Uranus passed in front of a faint star indicate that the planet is surrounded by rings of rocky fragments. The composition of the atmosphere of these outer planets has been deduced from spectroscopic measurements and appears to contain methane. Pluto appears as a small dot even in the largest telescope and is thought to be between a half and a tenth of the size of earth and may once have been a moon of Neptune. Pluto will soon move inside the orbit of Neptune making Neptune the farthest from the sun.

38 Pioneer 11 will reach Saturn and its rings in 1979. Until then the only pictures available will be from Earth based telescopes. The planet Uranus was unexpectedly found to have rings. When it passed in front of a star on 10 March 1977 observers were surprised to see that the star disappeared and reappeared five times before passing behind the planet.

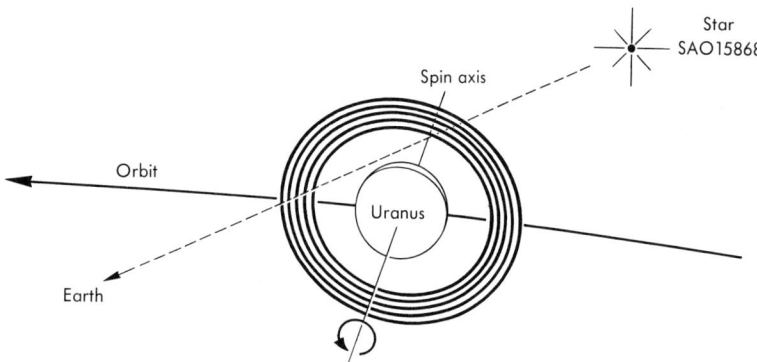

Planetary exploration spacecraft

The rapid development of unmanned planetary exploration since the 'fly-by' attempts of the early sixties is reflected in the achievements of the more recent Mariner, Pioneer, Viking and Venera spacecraft. Mariner 10 for example, returned many thousands of detailed pictures of the cratered moonlike surface of Mercury and the cloud tops of Venus in 1974. Both planets appear entirely featureless when viewed through Earth-based telescopes. In 1975, the Russian Venera 9 and 10 spacecraft penetrated the 20 kilometres thick cloud layer of Venus and landed on the surface. Despite the high surface temperature of 470°C, both craft survived long enough to transmit pictures of the rocky surface back to an orbiting craft which relayed the pictures back to Earth.

Pioneer 11 and 10 meanwhile, have returned pictures to Earth of the giant planet Jupiter from a distance of over 600,000,000 kilometres. The spacecraft were commanded from Earth and Pioneer 11 has responded to course corrections which will take it on a trajectory close to the mysterious rings around Saturn in 1979, six years after launch. Saturn is over 1,300,000,000 kilometres from Earth.

On 20 July 1976, the Viking 1 lander touched down on the surface of Mars, returning spectacular pictures of a desert-like landscape strewn with rocks. A second Viking landed at a site north west of Viking 1. The Viking spacecraft were launched in the August and September of 1975 as a combined orbiter and lander. Pictures from the orbiter's cameras were used to select a site for the landers.

The first man-made object to leave the solar system will be Pioneer 10. This dead remnant of our civilization will reach the present position of the star Aldebaran in 8,000,000 years time.

The external appearance of the spacecraft to some extent gives a clue to their function. The wing-like structures on Mariner 10 carry solar energy cells which generate electricity for the spacecraft whereas Pioneer 10 and 11 carry no such structures. We can immediately

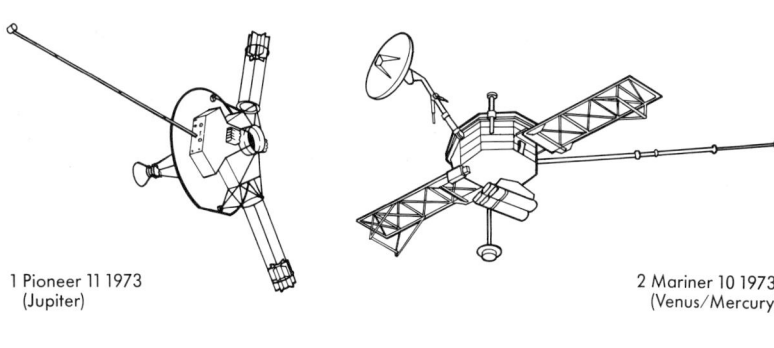

1 Pioneer 11 1973
(Jupiter)

2 Mariner 10 1973
(Venus/Mercury)

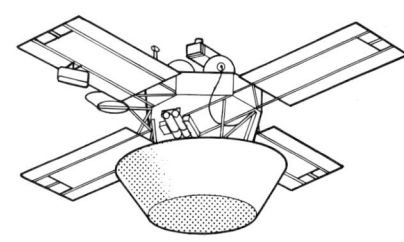

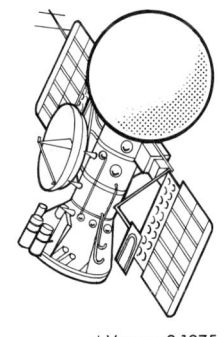

3 Viking 1975
(Mars)

4 Venera 9 1975
(Venus)

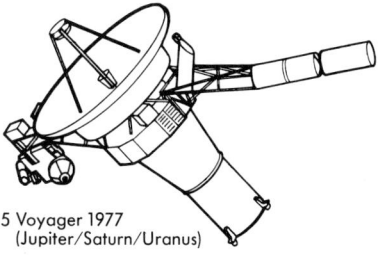

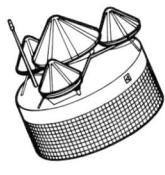

5 Voyager 1977
(Jupiter/Saturn/Uranus)

6 Pioneer Venus 1978
(Venus)

39 Planetary exploration spacecraft.

(23)

40 Communications are an essential part of planetary exploration both in transmitting commands to the spacecraft and in receiving data from it. The photo shows the tiny S-band antenna which was fitted to the Viking Lander to receive signals from Earth. The aerial in the background receives signals from the spacecraft.

41 The Pioneer-Venus spacecraft, due for launch in 1978, will carry four probes designed to penetrate the dense atmosphere of Venus. The photograph shows the four conical probes mounted on the main spacecraft. The probes will be released as the craft approaches the planet.

guess that Mariner 10 is designed to operate in the vicinity of one of the nearer planets to the Sun whereas Pioneer 10 and 11 must be designed to operate far from the Sun where solar energy is not sufficient to generate the electricity required. These are only superficial differences however and the major differences are given below.

Mariner class spacecraft

The standard Mariner body is an eight-sided structure 0.45 metres high and 1.39 metres across. Electronics compartments are built into each side. The spacecraft carries twin TV cameras and other instruments on a platform mounted below the craft. Launch weight of Mariner 10 was 503 kilograms. The spacecraft is stabilised by 3 small nitrogen gas jets, and referred to as 3 axis stabilisation. A hydrazine thruster motor is used for course correction. Electric power for the Mercury, Venus and Mars versions of Mariner was derived from solar cells designed to deliver 500 Watts in the vicinity of the planet.

1977 will see the launch of two Mariner craft designed to operate not only in the vicinity of Jupiter and Saturn but also near Uranus and Neptune 3,000,000,000 kilometres away. These craft (named Voyager) will derive their electric power from small nuclear generators similar to those used by Pioneer 10 and 11.

Pioneer class spacecraft

The Pioneer spacecraft are cheaper, lighter and generally simpler than the Mariner class and rely on spin rather than gas jets to stabilise the craft, referred to as spin stabilisation. Small gas jets are used occasionally for attitude control and course correction. Launch weight of Pioneer 11 was 259 kilograms. Telescopes on board Pioneers 10 and 11 used the spin of the spacecraft to scan the planet below and build up images which were then computer processed back on Earth to remove distortions due to the combined effect of the spacecraft movement and the planet's rotation.

Electric power for Pioneer 10 and 11 was produced by four small nuclear generators using the radioactive decay of plutonium 238 to generate heat which is then converted to electricity. 140 Watts of electrical power are generated in this way dropping to just over 100

(25)

Recent Planetary exploration missions

Spacecraft	Planet	Launch	Arrival	Comments
Mariner 9 USA	Mars	30.5.71	13.11.71	Over 7 000 pictures of Mars transmitted as the craft orbited the planet.
Pioneer 10 USA	Jupiter	3.3.72	4.12.73	Transmitted pictures of Jupiter's cloud tops as the craft flew by the planet.
Venera 8 USSR	Venus	6.4.73	22.7.72	Landed on Venus, transmitted data on radioactivity, sunlight penetration and windspeed.
Pioneer 11 USA	Jupiter & Saturn	6.4.73	3.12.74 (J) 9.79 (S)	Transmitted detailed pictures of Jupiter's cloud tops. Now heading for Saturn.
Mars 4 **Mars 5** USSR	Mars	21.7.73 25.7.73	1.74	Orbited Mars and transmitted pictures.
Mars 6 USSR	Mars	5.8.73	2.74	Landed on Mars but failed to operate.
Mars 7 USSR	Mars	9.8.73	–	Radio contact lost.
Mariner 10 USA	Venus & Mercury	3.11.73	5.2.74 (V) 29.3.74 (M)	Transmitted over 8 000 pictures of Mercury and Venus.
Venera 9 **Venera 10** USSR	Venus	8.6.75 14.6.75	22.10.75 25.10.75	Landed on Venus and transmitted the first ever pictures of the surface.
Viking 1 **Viking 2** USA	Mars	20.8.75 9.9.75	20.7.76 3.9.76	Many thousand detailed pictures of the planet both from orbit and from the surface were transmitted by the orbiters and landers.
*****Voyager 2** *****Voyager 1** USA	Jupiter & Saturn	8.77 9.77	1979 (J) 1981 (S) ? Uranus	Ambitious mission to the outer planet.
†Pioneer 12 **†Pioneer 13**	Venus	5.78 8.78	12.78	Spacecraft will carry probes intended to penetrate the atmosphere of Venus.

*previously named Mariner 12 and 11 respectively
†Pioneer Venus

(26)

Watts after 5 years. The generators are mounted on two 2.7 metre booms shown in the diagram.

In 1978 two Pioneer craft will be launched towards Venus, one to orbit the planet and the other to serve as a carrier for four probes which will penetrate the dense atmosphere. These craft generate electricity from solar cells mounted on the body.

Viking class spacecraft

The two Viking spacecraft launched towards Mars in 1975 comprised an orbiter based on the Mariner designs and a lander which weighed 576 kilograms and which separated after the craft arrived at Mars.

The lander was carried in a pressurised and hermetically sealed glass fibre shell beneath the orbiter. The shell was discarded during the lander's descent to the Martian surface. The lander, which derived electrical power from two radioisotope generators, carried two cameras and was equipped to perform various experiments including a search for life on Mars.

American craft are tracked by the NASA Deep Space Network of 64 metre diameter antennae located in America, Spain, South Africa and Australia. The giant Arecibo Radio Telescope whose 3000 metre dish is formed by perforated metal sheets suspended above a natural valley in Puerto Rico will be used to track and receive data from spacecraft at the limits of communication in the outer solar system.

The Russian Venera spacecraft launched towards Venus comprised an orbiter and lander which, unlike Viking, separated 48 hours before arrival at the planet. The Venera 9 and 10 were designed to withstand the high pressure and a temperature of over 470°C on the surface of Venus for only 30 minutes after being cooled to −10°C during the approach to the planet. In the event, Venera 9 functioned for 53 minutes and Venera 10 for 65 minutes, effectively doubling the scheduled operating time on the surface, to send back the first ever pictures from the surface of Venus.

The Russian Mars series of spacecraft, are pre-programmed before launch and cannot be commanded from Earth. The programme has achieved limited success but the landers have mysteriously suffered communications failures on touchdown.

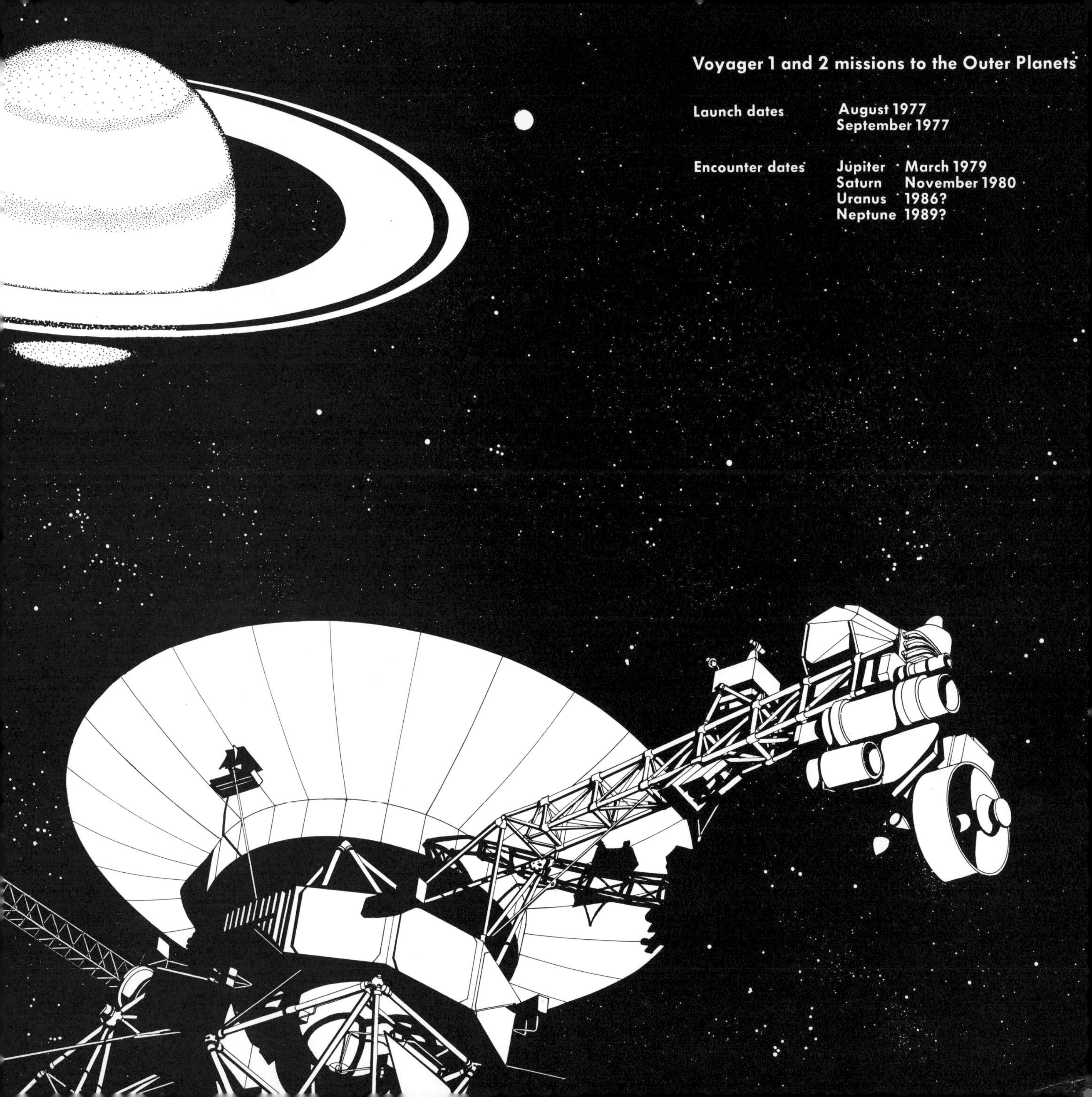

Further reading

GENERAL ASTRONOMY

Violent Universe Nigel Calder, BBC Publications.

Introductory Astronomy Pananides Addison Wesley.

New Horizons in Astronomy Brand & Maron, Freeman.

THE SOLAR SYSTEM

Earth Moon & Planets Whipple, Harvard Press.

The Solar System (Scientific American Book), Freeman.

The Comets Patrick Moore, Lutterworth Press.

PLANETARY EXPLORATION

Robot Explorers K Gatland, Blandford Press.

Pioneer Odyssey NASA publication SP-349 (NASA Scientific & Technical Information Office, Washington DC USA.)

MAGAZINES AND JOURNALS

Sky and Telescope (not generally available but may be purchased from the bookstall of the Old Royal Observatory, Greenwich).

New Scientist and *Scientific American*. These deal with many scientific topics and are generally available at bookstalls.

Printed in England for Her Majesty's Stationery Office
by The Soman-Wherry Press Limited, Norwich.
Dd 587511 K280 9/77